FLORA IN FOCUS
SEEDS
AND FRUITS

SMITHMARK

For this English language edition:
Todtri Productions Ltd., New York

This edition published in 1996 by SMITHMARK
Publishers, a division of U.S. Media Holdings,
Inc., 16 East 32nd Street, New York, NY 10016.

SMITHMARK books are available for bulk pur-
chase for sales promotion and premium use. For
details write or call the manager of special sales,
SMITHMARK Publishers, 16 East 32nd Street,
New York, NY 10016; (212) 532-6600.

Editions of this book will appear simultaneously
in France, Germany, Great Britain, Italy and
the Netherlands under auspices of Euredition bv,
Den Haag, Netherlands

Photographs:
PPWW Plant Pictures World Wide, Daan Smit
Text:
Nicky den Hartogh
Translation:
Tony Langham
Concept, design and editing:
Boris van Dobbenburgh
Typesetting:
Mark Dolk/Peter Verwey Grafische Produkties
Color separation:
Unifoto PTY LTD, Cape Town
Production:
Agora United Graphic Services bv, 's-Graveland
Printing and binding:
Egedsa, Sabadell, Spain

ISBN 0-8317-6122-9

INTRODUCTION

In the spring the feathery pappi of willows and poplars flutter down onto the fields like small snowflakes. Early in the year the dandelion seeds wait for a breeze to blow them on their widely stretched parachutes, to a new life. When the time is ripe, the spindle tree, yew tree and rowan tree enclose their seeds in a colorful jacket and the branches of the ash and maple tree are decked with winged fruit. The seed pods of the witchhazel pop open with the force of a small explosion, so that the seeds shoot out over a great distance like miniature rockets. The seeds of the alder are carried along on small aircushions by the current of the river.
All this activity is directed towards a single goal: the survival of the species.

The incredible reproductive drive which exists in the plant world is vitally important to man, because a large part of our daily food intake - varying from cereals, nuts and beans, to fruit - consists of fruit and seeds. Many animal species are wholly or partly dependent for their survival on the fruits of the fields and woods, and in turn, serve as a source of food for other animals. Fruit is an essential link in the eternal cycle of nature.
The fact that the beautiful way in which fruit is formed and seeds are dispersed seems completely unimportant in the light of millions of years of evolution, but this does not stop many people from being fascinated by the treasures which nature has to offer. The intriguing process starts at the moment that the flowers first open up. There cannot be a fruit without a flower. At least, this applies to the so-called flowering plants which cover a large part of the earth.

Well protected from the dangers of the outside world, the eggs (known as ovules in plants) are safely hidden away in the ovary, which forms part of the pistil - the female reproductive organ. In order to fertilize the eggs, (male) pollen is needed. This is produced by the stamens. From the time that the first flowering plants appeared on earth about a hundred million years ago, insects have played an increasingly important role in transporting pollen. Beetles, flies, bees and butterflies go from flower to flower and pass the pollen from the filaments at the top of the stamens to the often rather sticky stigmas at the top of the pistils. To attract them, the flowers produce brightly colored petals, secrete a sweet-smelling nectar, or assume the shape of an inviting female insect to which the male feels so attracted that he simply has to land on the flower.
And all this wondrous beauty serves a single purpose: the survival of the species.

When the pistil has accepted a grain of pollen, the fertilization takes place in the ovary. The egg develops into a seed, the ovary develops to become a fruit, surrounding one or more seeds. When other parts of the plant grow at the same time, such as the ovary wall, for example, this forms a false fruit, as in the case of a strawberry or fig.
The seed contains a germ from which the young plant is later produced. As this germ is not able to function independently while no green parts have developed, the seed also contains a greater or smaller amount of reserve nutrients.
In the soil where the fruit or seeds fall there is enormous competition for light, space and water. In the course of evolution, all sorts of ingenious systems have developed to safeguard the new generation. Some plants - pioneers in the plant world - specialize in colonizing barren wastes within a short time before the competition arises, and for this purpose they form large numbers of light seeds which are easily transported; other plants have seeds which are strong and contain a great deal of food, so that they can survive for a long time while they wait for a good opportunity to germinate.
The basic process is always the same, but there are infinite variations on the theme.

Coconuts are washed up on tropical coasts, sometimes after being transported in salt sea water for thousands of miles. Still they are able to germinate. The coconuts which are sold in shops are not fruit at all, but enormous seeds, surrounded by a rock-hard shell containing a large amount of coconut flesh. The fruit, which initially contained this seed, had a thick fibrous layer which greatly increased its capacity to float.
The four-to six-inch-long seed of the coconut palm is surpassed in size only by the "coco-de-mer" (Lodoicea maldivica) from the Seychelles, which has the largest seeds in the world.
Coconuts contain so much food for the journey and have such a large chance of germinating and developing, that it is not necessary for a coconut palm to produce many fruits.

In contrast, orchids produce countless seeds for every plant. This is essential, because they contain hardly any nutrients and can germinate only in a soil that is specially suitable for orchids. The seed of some species is as fine as dust, and so light that there may be 200,000 seeds to a fraction of an ounce. They are dispersed by the wind.

Airborne seeds must be light, and therefore contain few reserve nutrients, so that they do not usually retain their capacity to germinate for a very long time. Plants with lightweight seeds usually compensate for the short lifespan of the seed with an incredible over-production. It is often calculated that if all the tiny seeds produced by a single willow or poplar in the course of its life were to develop into mature trees, the world would be covered by a huge poplar or willow forest.

In order to increase the chances of dispersal, many plants surround their seeds with tasty, fleshy fruit, so that they form an attractive source of food for animals. In many cases, bright colors and tempting odors serve to attract the animals' attention. Birds peck colorful berries from branches. They digest the fruit, but the pip remains undigested and is usually deposited in a new place with some manure.

Mammals carry fruit with them, only eat half, and then leave the remains behind. In the tropics, fruit-eating hornbills, toucans and bats disperse the undigested seeds in their excrement; with their sensitive lips, chimpanzees are able to separate out the stones while enjoying the fruit.

Goldfinches batter at the heads of thistles so fanatically that the seeds fly all around. Parrots' beaks have evolved in such a way that they can crack hard nuts, and in the pine forests of the northern hemisphere, crossbills perform all sorts of tricks to remove the seeds from the hard scales of the cones.

The importance of safeguarding a new generation is not accompanied by such exuberant external displays in all plants. A great deal is concealed and takes place out of sight of the human eye, but it is no less fascinating for all that.

For example, an "antbread" - a barely visible part of a usually tiny seed - is of vital importance for the dispersal of plants. Ants consider this small appendix to be a delicious morsel. They drag the seed along to their nest, eat the nutritious "bread," and then lose interest and leave the seed. The pappi of goose grass, and large burrs attach themselves to people and animals by hooking onto clothes or fur with tiny barbs. Grains with long palea or pieces of the inflorescence are transported in the same way. In some grasses the palea respond to the moisture in the earth by stretching and turning, so that they work themselves into the ground without any help from outside.

The majority of the seed that is produced by plants is lost after being deposited on infertile soil. In addition, a significant proportion is eaten, but nature is prepared for this. Even if only a fraction - possibly just one in a million - of the seeds fall on good soil, they can germinate to produce fruit-bearing plants. This completes the cycle, and ensures that there will be new generations.

TRAGOPOGON PRATENSIS

When the flowerhead of the goat's-beard (Tragopogon pratensis), which is more than two inches across, forms seed, it is composed of a magnificent structure of numerous parachute seeds. The small "achenes," as the single-cell fruits are called, are ready to be dispersed by the wind. They are connected to the pappus of finely sprung, shiny hairs by a long, thin stem (beak). When the seeds have been blown away by the wind, the goat's-beard dies off. If the fruit did not set, for example, because the flower was prematurely mown down, the plant usually lives longer, and in this way has another chance to regenerate. Goat's-beard can be found above all on verges and on canal banks, by railway lines and rivers. The yellow flowers open in good weather, usually only in the morning. The plant belongs to the Compositae family. Like the dandelion, daisy and other members of this family, the "flower" is composed of a large number of tiny florets embedded so tightly together on the head that they appear to be a single flower.

TARAXACUM OFFICINALE

The bright yellow flowers and heads of the dandelion (Taraxacum officinale) adorn the grass and verges for a large part of the growing season. They grow at the end of a leafless, hollow stem, which contains a white sap. In the past, all sorts of children's games were played with the dandelion heads. For example, after blowing three times, the number of seeds left on the heads would indicate the number of children you would have. At the bottom of the thin stems of the pappi, there are oblong, ribbed brown achenes. When these are blown away by the wind, they leave small holes in the white head, where they were attached.

CLEMATIS VITALBA

Old Man's Beard (Clematis vitalba) attaches itself to bushes and the branches of trees with tendrils. The stems can grow up to a length of thirty feet and easily cover the distance from the top of one pine tree to another. In summer, this creeper has greenish-white flowers joined together in loose plumes. The calyx contains the stamens as well as a large number of pistils. Although the plumes of the flowers are quite attractive in themselves, the plant only really becomes striking when it produces fruit. While the seeds develop inside the ovary, the styles turn into long "beaks" which surround the tightly-packed brown achenes like elegant white tails. When the leaves have fallen from the tendrils, the magnificent downy pappi of the white-haired styles continue to adorn the branches until the depths of winter, and the fruit is eventually dispersed by the wind.

CLEMATIS VITALBA
SOLANUM DULCAMARA

The creeper-like stems of Old Man's Beard are supported by tree trunks, or spread so widely over the bushes that in autumn and winter a large part of the vegetation can be covered by a veil of white fluff. Often many other creepers can be found in the same place as Old Man's Beard, such as Woody Nightshade (Solanum dulcamara) shown here. Its oval, ripe berries have a bright orange color and are slightly poisonous. Its other name, "bittersweet," does not relate to the taste of the berries, but to that of the young stems which children used to chew in the old days instead of sweets.

CLEMATIS SPECIES

A close-up look at the feathery styles on the fruit of some clematis species shows that they look like the feathers of exotic birds.

PAPAVER SOMNIFERUM, "PAEONY FLOWER-ED SALMON" (flower)

PAPAVER SOMNIFERUM "PROLIFERUM" (fruit)
In principle, the female part of the flower - the pistil - consists of three parts: the stigmas on which the pollen lands, one or more styles, and the ovary where the eggs are fertilized. A number of plants, including poppies, have a large ovary above the flower, which takes up a striking position in the center of the flower. Because the poppy lacks a style, the stigma are placed on the broad top of the ovary. Later on, the ovary is transformed into a large pod with small holes along the upper edge which form when the seeds are ripe. When the poppies are blown in the wind, the seeds are gradually shaken out, like pepper from a pepper pot. The poppy seeds from the opium poppy (Papaver somniferum) are used to decorate bread and savory biscuits. The dried sap from the unripe capsules is raw opium, the raw material of morphine compounds.

PAPAVER RHOEAS
The common poppy (Papaver rhoeas), together with the charlock (Sinapis arvensis).

EUONYMUS PLANIPES

Euonymus planipes, an Asian spindle tree, has carmine pods containing seeds with an orange case.

PITTOSPORUM CRASSIFOLIUM

When the fruit of Pittosporum crassifolium has split open entirely, the black seeds remain lying on a sticky layer at the top of the open seed pod.

EUONYMUS EUROPAEUS
Many plants depend on birds for the dispersal of their seeds. To attract tits, thrushes and robins, the white seeds of the spindle tree (Euonymus europaeus)) are covered with an attractively colored fleshy seed covering. At first, they are completely enclosed by a four-celled pod. It is only when this opens in the autumn, and the orange coated seeds hang by a sort of umbilical cord under the flap of the pod covers, which are between a deep pink and purplish-red color, that the exceptional combination of colors is revealed. It is not advisable to eat the tempting fruit of this shrub, because it is very toxic to man, and there have also been reports of sheep, horses and goats being poisoned. In the past the orange-yellow substance from the seed covering was sometimes used to color butter.

CLEMATIS "BARBARA JACKMAN"
The striking spherical heads of
Clematis "Barbara Jackman," a garden
clematis with magnificent, large, pur-
plish-violet flowers, are formed by the
long, rolled-up beaks which adorn the
achenes.

POPULUS LASIOCARPA
The female catkin of the large-leaved poplar (Populus lasiocarpa)
from China, after it has finished flowering. The willow and the poplar,
which both belong to the willow family, form enormous quantities of
pappi, which flutter down like snow, covering the roads and fields with
a downy white layer. The small pods of the poplar develop in spring
on the catkins of the female trees. When they are ripe, they open up
under the pressure of the pappus, and the seed is then dispersed by the
wind. ▶

CLEMATIS "BARBARA JACKMAN"
Close-up of the pappus of the fruit of Clematis "Barbara Jackman."

Fluffy pappus, of the fruit of a clematis.

ACER SEMPERVIRENS
Branches of Acer sempervirens (syn. Acer creticum), a maple shrub, which has leaves in summer and winter, and comes from Greece and the eastern part of the Mediterranean, are adorned with magnificent red fruit with glabrous wings.
In spring, maples have small male, female or androgynous flowers. Self-pollination is avoided because the stamens usually appear at another time than the pistils.
All sorts of insects which collect nectar transfer pollen from one tree to another, and in this way contribute to pollination. In some species of maple the flowers are pollinated by the wind.

AILANTHUS ALTISSIMA
The tree of Heaven (Ailanthus altissima) is indigenous in China, but is also found in the wild in North America and Southern and Central Europe. Like the ash and the maple, the tree of heaven has very long fruit with glabrous wings. Together these form large plumes which are most striking when they turn from a yellowish-green to an orange-red in the autumn. The small seed in the middle of the more or less twisted wings is easy to see; it is connected to the outside edge with a sort of umbilical cord. Often part of the fruit remains hanging on the branches throughout the winter. Another part is dispersed and germinates while the wing is still attached to the seed. When the germinating root has anchored itself in the soil, and some small lateral roots have formed, two green seed cotyledons unfold. They look rather like leaves, but have a completely different shape from the true leaves. It is only at a later stage that the young plant develops the long feathered leaf which is characteristic of the tree of Heaven.

ACER GINNALA

Acer ginnala, a maple shrub which grows naturally over a large part of Asia, has keys which hang in compact bunches. Maples have split fruit, consisting of two halves. In each half there is a seed with a long sturdy wing. In the autumn the two halves of the fruit separate and flutter down, rotating like the propeller of a helicopter. In this way they can land so far away from the parent tree that they have a good chance of settling in a spot where the plant will have adequate light and space to be able to develop.

SHOREA COMPRESSA

The winged fruit of Shorea compressa constantly rotates on its own axis as it drifts down from the top of the tree to the forest floor.

A CACIA LONGIFOLIA
In plants belonging to the papilionaceae family, the seeds are in a row in a pod. Many fruits which grow in pods, including peas and beans, are important vegetables.
The pods of Acacia longifolia from Australia are woody, and about five inches long. When they are mature, they spring open along the front and back edges, and the red seeds are revealed and pecked out by birds.

B RACHYCHITON POPULNEUS
The attractive pod of Brachychiton populneus.

XANTHOCERAS SORBIFOLIUM

XANTHOCERAS SORBIFOLIUM
The hard pod surrounding the seeds of Xanthoceras sorbifolium from China is reminiscent of the shape of the husk of a horse chestnut.

EPILOBIUM HIRSUTUM

The great hairy willow herb (Epilobium hirsutum) grows on the banks of rivers, canals and on the shores of lakes. It is often the first plant to appear on a large scale on areas of land that have been drained or reclaimed. In summer, the plant is adorned with beautiful, purplish-red flowers at the tips of long, stem-like ovaries. Every plant produces many thousands of seeds. These are found in elongated pods, of which the walls gradually split open along the entire length of four seams. The four narrow parts curl outwards. The small light seeds have a parachute which is attached to two of the parts of the wall. When the fruit opens from top to bottom, the parachute attached between the two diverging parts spreads out so that it is caught by the wind. Once the hairy willow herb has become established, it covers more ground by putting out long underground root runners.

SETARIA ITALICA "MACROCHATEA"

Setaria italica "macrochatea" owes its English name, "foxtail," to the fluffy overhanging plumes of the ears. The ripe seeds are pecked from the plume by birds, and the plant is actually cultivated for birdseed, for example, in Italy. Setaria species were cultivated by man even in prehistoric times to produce strains which yielded more seed than that produced by wild grasses. However, they were crowded out by other cereals when agricultural methods were intensified, and are now mainly important in areas where the soil is dry and poor in nutrients.

PSEUDOTSUGA MENZIESII
A long time ago the tall Douglas fir from North America was imported into Europe, where it is now an important tree in forestry. This conifer can easily be identified by its characteristic hanging cones, in which the trident bracts appear between the seed scales like a festive fringe.

QUERCUS ROBUR
At the start of their development, acorns are completely enclosed by a scaly cup, which only covers the foot of the fruit when it has matured. When they fall, the acorns land near to the parent tree because they are very heavy. The fact that young oak trees also grow in spots further away, is due, amongst other things, to creatures such as jays and squirrels, which collect the acorns for their winter stores. Although only a small number of the acorns are lost by these creatures, and an even smaller proportion ultimately falls on fertile soil, there are enough fruits which germinate, and thus keep the population of oak trees at a constant level. ▶

NEPHELIUM LAPPACEUM

In the tropics, rambutan, achiote and sirsak are valuable consumer crops. The fruits of the rambutan (Nephelium lappaceum) are deliciously juicy. The taste and structure of the transparent white flesh of the fruit underneath its soft prickly skin resembles the well-known lychee. In Malaysia and Indonesia, where the rambutan is indigenous, it is cultivated as a fruit tree. The rambutan orchards are beautiful when the branches are heavily laden with red or yellow fruit.

BIXA ORELLANA

The brightly colored seed covers of the achiote, or lipstick tree (Bixa orellana) produce a virtually tasteless and odorless red dye which is used to color butter, cheese and soups, and has become an important commodity for export in many countries. For the Indians in the original natural habitat (tropical America), the tree was one of the most important sources of the red dye which they used to paint their bodies.

ANNONA MURICATA
The shell of the sirsak, or sour sop (Annona muricata) from Ecuador and Peru, is covered with soft thorns. Underneath there are a large number of segments of fruit with creamy white sour-sweet tasting flesh around a brownish black stone.

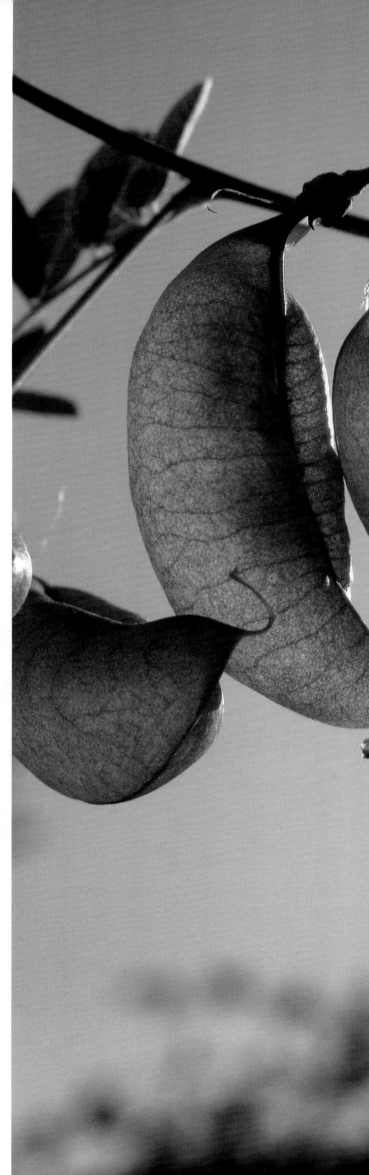

COLUTEA ARBORESCENS
The bladder senna (Colutea arborescens) has transparent, bladder-shaped, swollen pods after it has finished flowering. ▶

GOMPHOCARPUS FRUTICOSUS
The large balloon-shaped fruits of Gomphocarpus fruticosus, a herb-like plant with narrow leaves. It is indigenous in tropical and South Africa, and has become widespread in America.

SOLANUM MAMMOSUM

The nightshade family (Solonaceae) includes plants with the most diverse shapes of fruit, from peppers and eggplant to the small berries of bittersweet. Perhaps the most curious fruit of all is produced by Solanum mammosum. This small shrub comes from tropical America, and is known in English as "nipple fruit," because of the characteristic growth of the fruit.

A NACARDIUM OCCIDENTALE

Fruit formation in the cashew nut (Anacardium occidentale). The kidney-shaped nut is attached to a fleshy, greatly swollen yellow or red stem. Peeled and roasted cashew nuts are popular all over the world. The fleshy false fruit on which the nut grows ("cashew apple") is edible, but does not taste as good as its fragrance suggests.

BERBERIS JULIANAE

The beautiful ripe berries of Berberis julianae appear in autumn, and hang on the branches for a long time. This species of sour berry comes from the Chinese province of Hupeh. It was introduced into Europe in 1900, and has been planted on a large scale since that time in parks, green belt areas and gardens.

SYMPLOCOS PANICULATA

Symplocos paniculata, a rich fruit-bearing shrub which comes from Asia, is also known as the sapphire berry, because of the beautiful blue color of the fruit. ▶

DIPSACUS FULLONUM

Fuller's teasel (Dipsacus fullonum) is a striking plant because of the large flowering heads at the ends of the tall, spiny stems. The flowering head is composed of a large number of florets, of which the bracts end in a stiff spine. The narrow, enveloping leaves at the base of the inflorescence curve round elegantly. The photograph here shows the beginning of the flowering period, because it is known that the fuller's teasel opens up first, halfway up to the flower. When flowers appear at the top and the base, the strip in the middle of the flowering head is already bare.

At the beginning of its life, the large fuller's teasel only forms a leaf rosette which remains pressed down against the soil for one or two winters. Only then does it develop the tall flowering stem. The flowers are pollinated by bees, flies and butterflies. As soon as the squarish false fruit, which have grooves, are ripe, the fuller's teasel's task is complete. The plant slowly dies away, but the stems and dry flower heads continue to adorn the surrounding area throughout the winter. When the flowerheads have died off, they retain their shape for a long time. Another teasel (Dipsacus sativus), which has hook-shaped, curved straw bracts, was used in the past for carding wool.

DIPSACUS SILVESTRIS

In Dipsacus silvestris, a teasel which is found in Southern and Central Europe, small plants develop from the seeds while they are still in the flowering head.

PTILOSTEMON AFER
The ivory-colored fluff of the "ivory thistle," which is indigenous in the Balkans and Asia Minor.

CIRSIUM ERIOPHORUM
When the tuft of purplish-pink tubular flowers has disappeared, the thick, spherical flowerhead of the woolly thistle (Cirsium eriophorum) is crowned with white fluff. The fruit of the thistle connected to this is a nut: a single cell, dry fruit which does not pop open, and has a close-fitting seed wall. The fruit of the woolly thistle is rich in oil, and very popular with songbirds such as the goldfinch or redcap, whose beaks are suitable for separating the nuts from the flower. The fluff of the thistles often comes away easily, and slowly drifts over the countryside in large flakes.

SORBUS AUCUPARIA

Bunches of fruit of the common rowanberry (Sorbus aucuparia). Rowanberries have a bitter taste, and are slightly poisonous for humans in an unprocessed form. They are very popular with song thrushes and starlings, and not long after the birds discover the ripe, bright orange fruit, a large part of the bunch will usually be pecked away. Despite the fact that they contribute considerably to dispersing the seeds, nurseries and gardeners consider this to be very unfortunate. In fact, the varieties of rowanberry which birds ignore because they are not interested, are the most valuable as fruit-bearing decorative trees in gardens and parks.

SORBUS CASHMIRIANA

Sorbus cashmiriana, from Kashmir and Afghanistan, is one of the most charming rowanberries with its elegant open crown, full of white fruit.

CHAMAEDOREA TEPEJILOTE

The deep black, ripe fruit of Chamaedorea tepejilote, a palm tree which grows naturally from Colombia to Mexico, is the size of a small olive. The size and appearance of the fruit of palm trees can vary considerably. For example, the fruit of the coconut palm has a thick, fibrous outer layer covered by a smooth skin. The rock-hard shell which surrounds the four- to six-inch-long seed, is underneath the fibrous layer. Confusingly, both the fruit and the seed have the name coconut. A date has soft flesh surrounding the stone, while the edible fruit of the salak is protected by a hard, shiny shell, consisting of overlapping scales. Every fruit, no matter how different, serves to keep the seed in optimum condition. Like the coconut palm, date palm and salak, Chamaedorea is cultivated, though in this case it is not the fruit which is consumed, but the unopened male flowers.

OCHNA SERRULATA

Ochna serrulata bears striking composite fruits, uniting a number of berry-like stone fruit or drupes. Initially these are green, and when they are ripe, they are a shiny black. They contrast beautifully with the bright-red crown of pointed sepals, which remain on the plant after the other petals have dropped off. This small shrub grows in tropical Africa.

RHIPSALIS MESEMBRYANTHEMOIDES

The berries which adorn the leafy segmented stems of Rhipsalis mesembryanthemoides, a cactus which originates from Brazil, are almost transparent. The majority of Rhipsalis varieties, also known as coral cacti, are epiphytic plants. They establish themselves in trees or narrow cracks in rocks, where they sometimes form hanging branches which can be seven to ten feet long. The berries are eaten by birds, and in this way the seeds can fall in the most impossible places. Rhipsalis is one of the few cacti with a natural habitat which is not restricted to America. For example, Rhipsalis baccifera, has been found in Ceylon since time immemorial. However, this does not prove that the origin of cacti should not be exclusively considered to be in America; it is presumed that Rhipsalis has also found its way to other parts of the world with the help of birds and sailing vessels.

NELUMBO LUTEA

The pale petals and golden-yellow stamens of the yellow lotus (Nelumbo lutea) form a crown surrounding the curiously shaped, broad and flat center. The flowers of the lotus are supported by sturdy stems, so they appear above the surface of the water. Most of the single-seed fruits are sunk in the center and only the spherical tops protrude above it.

NELUMBO LUTEA

When the seed ripens, the head of the yellow lotus becomes so heavy that it hangs down. The heavy seeds fall into the water and sink down to the muddy bed under their own weight. They can remain there for a long time without losing the capacity to germinate. Actually, the lotus does not depend entirely on its seed, because the plant can also be propagated with the help of rhizomes. In the waters of its natural habitat in the United States, the lotus was once so common that it was an obstacle to shipping, and it was considered necessary to combat this problem. This had the result that the charming water plant has gradually become extremely rare in the wild.

LEYCESTERIA FORMOSA

Leycesteria formosa, from China and the Himalayas, belongs to the honeysuckle family, which includes a large number of shrubs which bear a lot of fruit, including the snowberry, elderberry, guelder rose, and of course, honeysuckle. In the Leycesteria it is not only the shiny fruit that is attractive, but in addition, the beautifully harmonizing, purplish bracts contribute to the shrub's beauty in autumn. The juicy fruit contains many seeds and is particularly popular with pheasants.

◀

LONICERA CAPRIFOLIA "INGA"

The common honeysuckle (Lonicera caprifolia) has been cultivated since ancient times, and is very common as a garden plant. Originally this well-known climbing plant came from Southeast Europe, the Caucasus, Asia Minor and Northern Iran. In summer, wonderfully scented flowers blossom at the ends of the twigs. Between the bunches of flowers there are always two fused, dish-shaped leaves. At a later stage, the colorful ripe berries in the middle of the round pairs of leaves are arranged as though they are served on a dish.

RUMEX VESICARIUS

The greenish or pinky-red petals of sorrel develop into a triangular case which surrounds the fruit. In several varieties of sorrel these cases become so broad at the edges that they serve to catch the wind, and in this way help to disperse the seed, This sort of development can be clearly seen, for example, in Rumex vesicarius, a variety of sorrel which originated in Greece, North Africa and Asia. ▶

CHENOPODIUM QUINOA

The large plumes of Chenopodium quinoa are found in countless different colors. In South America this tall goosefoot has been cultivated since ancient times, and is grown in particular in the Andes to supply seeds. The seeds are eaten boiled, or processed to make flour for baking bread.

EUPHORBIA CANARIENSIS
The three-celled pods of Euphorbia canariensis ripen from August. Initially they are green; later they turn a purplish-red color. When they are fully grown, they pop open with such force that the seeds are dispersed over a great distance.

EUPHORBIA CANARIENSIS
The pillars of this spurge plant rise up high above the dry grasses in the southern part of Tenerife (Canary Islands), where several other succulent varieties are indigenous in addition to Euphorbia canariensis. The magnificent purplish-red flowers grow in serried rows at the tops of the pillars.

ARONIA MELANOCARPA

The black chokeberry (Aronia melanocarpa) in the autumn. This shrub is closely related to the apple and pear tree, and like those important fruits, belongs to the rose family. Although the small fruits look like berries, they have the same structure as apples, with a miniature core and pips.

A RONIA MELANOCARPA
The natural habitat of Aronia, the black chokeberry, is
in North America. The shrub is planted in parks and gardens for its beautiful autumn colors and shiny, dark purple
bunches of fruit.

FICUS SYCOMORUS

The figs of Ficus sycomorus grow close together on the trunk and branches. The sweet fruit of the sycamore fig is eaten in tropical Africa and Asia, but it is smaller and of poorer quality than the common fig (Ficus caria). Figs are actually false fruit: the real fruit is the countless small "pips" inside the fig.

FICUS FISTULOSA

The pear-shaped object which is thought of as the fruit of the fig tree is in fact a curiously formed, fleshy shell. When the tree blossoms, it forms a hollow space in which hundreds of florets can develop. Gall wasps often penetrate the hollow through the narrow opening at the top of the shell to lay their eggs, and this results in pollination. There appears to be a special gall wasp for each of the many ficus varieties which depend on insects to be pollinated. The true fruit is produced by the fertilized florets. These "pips" are very small, so that they are eaten with the fig almost unnoticed.

In deciduous trees and other angio-
sperms the seed develops from an egg in
the female ovary, but in conifers (which
have cones) the seeds are unprotected in
the axilla of the scales of the cones.
Conifers are therefore known as gymno-
sperms. The scales composing the cones
are usually woody, as in evergreens such
as the larch, fir trees, spruce and silver fir.
In cypresses they are initially spherical
and less obviously like cones, while the
fleshy cone of the juniper berry looks like
a berry. Usually the seeds are mature
only in the second autumn following fertilization. Some conifers have winged
seeds so that they are dispersed over
large distances.

CUPRESSUS MACROCARPA
Young cones of Cupressus macrocarpa, a cypress from California.

JUNIPERUS COMMUNIS
Fleshy berries of the common juniper
(Juniperus communis).

-48-

PSEUDOLARIX AMABILIS
The cone scales of the golden larch (Pseudolarix amabilis) compose beautiful, initially light-green and later woody brown clusters.

PINUS MUGO
The pine cone of Pinus mugo, the mountain pine. The white deposits on the surrounding needles are the fruiting bodies of a fungus.

ATROPA BELLADONNA

Atropos, one of the three Fates in Greek mythology, cuts off the thread of life when the time has come. She symbolizes the inescapable, the inevitable, death. The fact that Atropa belladonna (also known as deadly nightshade or dwale) is named after her is significant in itself. The shiny black "cherries," crowned with star-shaped calyxes, contain an extremely poisonous juice which played an important part in the history of feuds and wars. Poisoning by deadly nightshade is accompanied by confusion and hallucinations and can result in death if a large quantity is consumed. The natural habitat of deadly nightshade in Europe extends from the South to Central England in the northwest. The plant was introduced in Denmark and Sweden, where it was not indigenous, and it became widespread. In the Netherlands it can be found occasionally on the wooded slopes of South Limburg. Birds eat the berries without any negative effect, and disperse the seeds in their excrement in open deciduous woodland and scrubland.

HELICONIA PENDULA

The combination of ripe and unripe fruit on the elegant hanging inflorescence of Heliconia pendula when it has finished flowering, and the red bracts covering them, produce a beautiful effect of color. At the top of the inflorescence the bracts of this South American plant can grow to a length of one to one and a half feet. ▶

EUPHORBIA CHARACIAS

Herb-like spurge varieties such as Euphorbia characias and Euphorbia rigida are very different from, for example, Euphorbia canariensis, which looks more like a cactus. However, there are strong resemblances between Euphorbias as regards the structure of the flower and development of the fruit. It is not the flowers which are the most striking part of the inflorescence, but the bracts. Their color varies from green and yellow to bright red, depending on the variety. The actual flowers are strongly reduced, and are joined together in a cup (cyathium) containing the stamens and a pistil with three styles. To prevent self-pollination, the pistillate flower develops before the stamen flowers. Flies are particularly attracted by the fragrant sweet nectar secreted by four or five honey glands on the inside of the cup. These ensure cross-pollination. While the pistillate flower develops, the stem lengthens until the whole inflorescence with the three styles at the top hangs out of the cup.

EUPHORBIA RIGIDA

Following pollination, the inflorescences of Euphorbias develop into three-celled pods. When they are ripe, they pop open along three seams with such force that the seeds are ejected away from the parent plant. In addition, ants fre-quently help to disperse several varieties of euphorbia. They carry the seeds because they have an oily appendix (the so-called "antbread") which ants consider a delicacy.

Grasses are pollinated by the wind and therefore do not need colorful stalks to attract insects. On the contrary, pollination by the wind is facilitated because there are no petals. The beauty of flowering grasses is determined by the shiny glume surrounding the flowers, the palea in which the anthers end, or the silk-like hairs in the inflorescence.

Almost all grasses have androgynous flowers with stamens which produce pollen and a pistil to receive the pollen. When the time is ripe, the stamens soon grow to such a length that they dangle outside the flowers and are exposed to the wind. In some varieties of grass, this process takes place within a quarter of an hour. Large quantities of pollen are necessary to maximize the chance of pollination. For rye, a count has shown that for every fertilized egg cell, no fewer than 57,000 grains of pollen are produced.

When the fertilization has taken place, the inflorescence develops to become a fruit: the grain of cereal. This contains seed and often remains joined to the palea. The grains are dispersed by the wind, by animals, or when the grass grows on banks or shores, by water.

TRIFOLIUM STELLATUM

Trifolium stellatum (or "Star clover") grows on the stony ground of the southern Mediterranean. The inflorescence is like a head containing a large number of florets. When the fruit sets, the calyxes develop into beautiful, pinkish-red pointed stars, creating a sophisticated color combination with the white pappi.

The ripe ears on stalks of grass which have turned yellow contrast vividly with the background of the Argentine landscape plunged in dark shadows. ▶

CORTADERIA SELLOANA

Pampas grass (Cortaderia selloana) grows taller than the height of a man, and, as its name indicates, it is found on the vast, treeless grassy plains of Argentina and Uruguay. Pampas grass is an exception among the other varieties of grass. There are plants which only have male flowers, and plants with only female flowers, while in most other grasses the male and female characteristics are combined in every flower. At the fruit stage, female pampas grass is more charming than the male equivalent, because the large plumes at the ends of the stalks are less compact and more vividly colored.

RHIPSALIS BACCIFERA
The milky white transparent berries on the long, cylindrical hanging stems of the mistletoe cactus (Rhipsalis baccifera) resemble the fruit of the mistletoe, hence its name.

SOPHORA JAPONICA
Despite its name, the Japanese Pagoda tree or honey tree (Sophora japonica) is not indigenous in Japan, though it is in China and Korea. The tree is at least thirty years old when the first flowers appear on the branches. Pods develop when the trees have blossomed, as is characteristic of members of the papilionaceae. The pods of the Japanese Pagoda tree are four to six inches long, and are reminiscent of strings of beads because the pods are indented between the seeds.

PARKIA SPECIOSA
The pods of Parkia speciosa - the peteh bean - grow to a length of thirteen to fifteen inches. They contain edible seeds which are prepared in all sorts of ways, particularly on the Indonesian island of Java, where they are amongst the most highly valued vegetables. ▶

LONICERA PILEATA

Lonicera pileata, a low-growing, leafy shrub from Central and Western China, unlike many related plants with large flowers - the climbing honeysuckles - produces small, greenish-yellow flowers. Although they are not particularly attractive, they are very popular with bees, who are attracted by the strong fragrance of the flowers. When the shrub has blossomed, it is covered in countless transparent berries.

LONICERA XYLOSTEUM

The poisonous berries of fly honeysuckle (Lonicera xylosteum) are arranged in pairs on a common stem. Because of the lack of space, one of the two fruits is sometimes clearly less developed. They are dispersed by birds and germinate in light, deciduous woodland and thickets in chalky soil.

TAXUS BACCATA

Not only the twigs, but also the bark and seeds of Taxus baccata (the yew tree) are deadly poisonous. The large black seed is surrounded by a bright red seed cover which is not completely closed at the top. The yew berries, as the Taxus fruit is known, are eaten by birds. They only digest the fleshy, non-toxic seed cover. The seed is excreted, completely undigested, and germinates when it falls in soil which is suitable for Taxus plants.

TAXUS BACCATA LUTEA

The berries of Taxus baccata "lutea" are different in color from those of the common yew tree.

TRIOSTEUM PINNATIFIDUM

Triosteum pinnatifidum is a member of the honeysuckle family which bears a great deal of fruit. Its natural habitat covers Central and Western China.

TRIOSTEUM ERYTHROCARPUM
The color of the berry of this Triosteum variety is expressed
in its name. Erythrocarpum means "with red fruit."

CYCAS REVOLUTA

There are long, stiff leaves up to six feet long which surround the enormous orange-red seeds at the top of the stem of Cycas revoluta, a palm fern from Southeast Asia and Japan. Palm ferns are amongst the oldest seed plants in the world. They flourished particularly at the time when the earth was covered by gigantic horsetails and tree ferns, about 225 - 65 million years ago. Discoveries of fossil remnants have shown that they have barely changed at all over millions of years.

Like conifers, palm ferns are "naked-seeded," the class of gymnospermae. In Cycas, the naked egg cells of the female plants are placed at the edge of the feathered, leafy organs. The enormous flowering cone on the male plant produces huge quantities of pollen. When they are ripe, the rather flat seeds are a bright vermilion color, almost one and a half inches long.

CYCAS REVOLUTA
(Female plant)

CYCAS REVOLUTA
(Male plant)

GARDENIA THUNBERGII

G The large, rather woody fruit of Gardenia thunbergii, a South African evergreen shrub, sometimes remain on the stiff, greyish-white branches for years on end. They are eaten by antelope. The digestive juices of these ruminants do break down the tough fruit, but the seeds are not damaged and can still germinate after being excreted.

KIGELIA AFRICANA

K The fruit of Kigelia Africana, the sausage-fruit tree look like enormous liverwursts. They hang on exceptionally long stems, can grow to a length of seventeen inches, and weigh between eleven and fifteen pounds. They do not open after falling off the tree, and the fruit has to rot before the seeds are free to germinate. ▶

SCORPIURUS VERMICULATUS

Scorpiurus vermiculatus - the "Scorpion tail" - is an annual plant from Southwest Europe. After the yellow butterfly flowers have blossomed, the stems have curiously shaped pods, rolled up like a scorpion's tail. In addition, the surface is covered with burrs, so that the pod looks like a rolled-up caterpillar or centipede.

TULIPA
The seed pods of the tulip.

HELIANTHUS ANNUUS
The endless fields where sunflowers are cultivated for their oil-rich seeds provide a magnificent view. The large flower heads always turn to face the sun. Colorful outer florets spread out in a crown around the heart of inner florets, where the sunflower seeds develop at a later stage. ▶

HELIANTHUS ANNUUS (seeds)
Sunflower seeds ripen packed together in the heart of the sun-flower (Helianthus annuus). They have a high oil content and are used on a large scale as the raw material for sun-flower oil. The sunflower originated in South America, where it was cultivated even before the Spaniards first arrived, and is amongst the first plants to be brought to Europe from the New World.

SORGHUM BICOLOR

A ripe ear of Sorghum bicolor or guinea corn can contain up to 3,000 seeds. Like barley, rye, maize and wheat, sorghum is one of the grasses which are cultivated as a food crop. At a time when people still had a nomadic existence, they collected the protein-rich seeds of wild grasses. In the course of human evolution the plants which naturally have a richer fruit than other varieties were selected and cultivated, a process which continued over many centuries until the grasses barely resemble the original varieties and are completely adapted to human requirements. The inflorescence of the grasses is composed of a number of ears in which several minuscule flowers grow together. The flowers and ears are surrounded by the palea. The ears are joined by a central axis and jointly form a rigid ear, or flowering stem. In the natural state the ears with the ripe seeds often break off easily, together with part of the stem. This helps to disperse the seeds because the ears easily become stuck in an animal's fur, so that the seeds are transported to distant places. If the ears break off easily when they ripen, a lot of the grain is lost before and during the harvest. This is why this characteristic which occurs in nature is no longer found in cultivated varieties.

TRITICUM DICOCCUM

Triticum dicoccum is amongst the oldest cultivated varieties of wheat and is one of the ancestors of modern varieties. The origins of cultivated grasses go back more than 6,000, and perhaps even 10,000 or 17,000 years. Following a process of selection lasting thousands of years, it is estimated that there are now 10,000 different varieties, the grains of which are used for the preparation of bread, spaghetti, gin, beer, semolina and groats, depending on their qualities. There are some varieties with long palea, as in the case of Triticum dicoccum ("bearded" varieties), but there are also varieties without. The main part of the grain consists of the endosperm, rich in protein, which was not originally intended for human consumption, but serves to feed the germ (embryo) until it has developed as a plant and can function independently. Wheat is a particularly suitable grain for making bread, because the grains contain a great deal of sticky protein or gluten, which allows the dough to rise.

PANDANUS EDULIS
The fruit of Pandanus edulis, which resembles the pineapple, is eaten in Madagascar. It is composed of a number of angular orange segments.

NERTERA GRANADENSIS
In relation to the small size of this low-growing ground cover plant, the brightly colored, berry-like fruit is extremely large. The Coral berry (Nertera granadensis) grows in mountainous regions of the Southern hemisphere.

ARUM ITALICUM

The fleshy berries of the Italian arum lily (Arum italicum) are tightly packed together in a spike at the top of the sturdy flower stems. In contrast with the charming appearance of this flower, the taste of the fruit is positively disgusting. This is just as well, because like the splendid arrow-shaped leaves (not shown in the photograph) and the starchy rhizomes, they are rather poisonous. The spotted arum lily is indigenous in Southern and Western Europe and grows as far north as Southern England. It is planted in gardens as a decorative plant.

URTICA PILULIFERA VAR. DODARTII

Seen close up, the fruit of the most diverse plants are indescribably beautiful. The long-stemmed, compact fruit of Urtica pilulifera, is about one third of an inch across. This stinging nettle has poisonous, stinging hairs. The fruit contains shiny black seeds.

ARALIA RACEMOSA

The composite bunches of fruit of Aralia racemosa clearly reveal its relationship with the better known ivy plant. The two plants belong to the same family. In Aralia racemosa, which is indigenous in North America, the umbels in which the berry-like drupes are arranged, form a large, striking plume.

INDEX